# Tales of Modern Alchemy

## 3D Printing in Metals and Electronics

# Table of Contents

# Chapter 1. Introduction

Special Report: Tales of Modern Alchemy: 3D Printing in Metals and Electronics

In our world where science steadily punctuates the unbelievable into reality, we welcome you to join us on a fascinating journey with this Special Report. The Tales of Modern Alchemy: 3D Printing in Metals and Electronics bridges the realms of fantasy and reality, uncovering the transformative power of 3D printing. With a pragmatic and accessible approach, we delve into the principles and practices of using this breakthrough technology in metal and electronic fabrication. This report caters to professionals, enthusiasts, innovators, and dreamers alike interested in understanding the technical nuances, the potential societal impacts, and future implications of this exciting area. Let us demystify the often complex world of 3D printing together, as we explore and appreciate the modern alchemy reshaping our lives today. Secure your copy of this special report and down-to-earth guide to this pivotal science, unveiling unique opportunities for innovation and growth.

# Chapter 2. Unveiling 3D Printing: A Modern Alchemic Journey

3D printing, often referred to as additive manufacturing, is fundamentally altering the way we design, produce, and distribute objects, from small everyday household items to specialized components used in aerospace and medical industries, analogous to how a modern-day alchemist would. This riveting realm is transforming mere resources into outputs of increased complexity and functionality, closely reminiscent of the legendary practices of old-age alchemists, albeit without the mystical elements.

## 2.1. Introduction to 3D Printing

The concept of 3D printing is simple yet powerful. It refers to a process of creating three-dimensional solid objects from a computer-aided design (CAD) model by depositing successive layers of material. It enables the creation of complex, custom and intricate designs with ease, as each layer bonds to the previous one.

Unlike conventional manufacturing processes, which involve cutting away excess materials to construct a product, or casting and forging methods for mass production, 3D printing is an additive process. There are a multitude of available materials to fulfill distinctive roles - plastics for prototype design, metals for high-strength parts, and ceramics for specialized applications.

## 2.2. History and Evolution of 3D Printing

The concept of 3D printing was born in the 1980s, with the pioneering technology named 'stereolithography' invented by Chuck Hull. The following decades witnessed rapid development and evolution in this field. Today, it encompasses multiple techniques such as selective laser sintering (SLS), fused deposition modeling (FDM), and direct metal laser sintering (DMLS), to name a few. While the fundamental principle of layering materials remains, the precision, accuracy and usability of these technologies have evolved dramatically.

## 2.3. Principles of 3D Printing

The process of 3D printing starts with a digital design created in CAD software. This model is then sliced into thin layers in a process called slicing. Once the slicing is complete, the printer can start depositing or fusing materials layer by layer to create the physical object. The intricacies of this process can be adjusted in the software, allowing for intricate designs, precise dimensions, varying densities, and multifunctional properties.

## 2.4. 3D Printing in Metals

Metal additive manufacturing represents a leap in the world of 3D printing, lending it the tremendous power to create robust, highly customized structures that outperform traditional manufacturing methods. Materials like titanium, stainless steel, aluminium, and even precious metals such as gold and silver can be used.

This process combines fine metallic powders and a focused laser or electron beam to melt and fuse particles layer by layer, resulting in highly-durable, intricate designs. The size of the particles in the powder affects the accuracy of print, with finer particles allowing for greater detail.

It's not only about the physical characteristics of the end product, either. The ability to print with conductive materials opens up a world of electronic functionality that conventional manufacturing processes struggle to match. Integrated circuits, antennae, sensors - all can be printed with remarkable precision and complexity.

## 2.5. 3D Printing in Electronics

The application of 3D printing extends beyond tangible products. We can now print electronic components, ushering us into an era of printed electronics. This is done by inserting or printing conductive inks or materials, where the CAD design dictates. These are typically silver-, copper-, or carbon-based inks capable of carrying a current.

Imagine a future where you could print a fully-functional circuit board, complete with capacitors, resistors, and integrated circuits. This can pave the way for customizable electronics, reshaping the electronic industry in ways we have never seen before.

## 2.6. The Societal Impacts and Future Implications

These manufacturing breakthroughs promise to revitalize production practices, disrupt industries, and even reshape global economies. The capacity to produce customized, complex products in-house, in small quantities, and without expensive tooling, holds the potential for profound societal change. Moreover, it could significantly reduce environmental impact by minimizing waste generation, reducing transportation, and promoting sustainable material usage.

A world equipped with 3D printers in every home and every industry is not a distant fantasy, but a developing reality. However, challenges still exist. The legal landscape concerning intellectual property rights in the age of easily replicable items is of concern. The need to further reduce costs, improve speed and reliability, and develop new materials are equally pressing.

But that's part of the excitement. After all, what is alchemy without a sense of mystery, challenge, and discovery? As we continue our journey in understanding and mastering this modern alchemy of 3D printing, we are writing a new chapter in the annals of human creativity and invention.

Despite these challenges, the new scientific and technological frontier of 3D printing is undeniably upon us. Its ultimate potential remains largely untapped and it's up to us, the modern-day alchemists, to explore, experiment, and extract the immense value it promises to offer.

# Chapter 3. Understanding the Science: From Concept to Creation

To wholly grasp the science of 3D printing in metals and electronics, it is crucial to first understand the underlying principles of additive manufacturing. Unlike traditional manufacturing methods, which involve the elimination or deformation of material to create a product, additive manufacturing, or 3D printing, works by layering materials until the intended product takes shape.

## 3.1. Additive Manufacturing Fundamentals

At the heart of any 3D printer is a simple guiding principle: creating a three-dimensional object layer-by-layer, guided by a digital design. The 3D printer software slices the design into hundreds or even thousands of horizontal layers. These layers are then printed one over the other until the entire object has been formed.

Metal 3D printing, also known as metal additive manufacturing, employs these principles but uses metal powders or filaments instead. On the other hand, electronics 3D printing, or 3D printed electronics (3DPE), moves a step further, integrating functional materials into additive manufacturing to create electronic components.

## 3.2. Metal 3D Printing

Metal 3D printing begins with a thin layer of metal powder spread over a build plate. A laser or an electron beam, guided by the preprogrammed design file, precisely melts the metal powder, fusing it together and onto the build area. Once the layer is completed, the build plate moves down slightly, and another layer of metal powder is spread out.

This process continues until the design is fully realized, after which the object is allowed to cool and then removed from the printer. Post-processing, involving heat and solvents, may then be required to improve part quality or remove residual powders. The exact process, however, may vary depending upon the metal printing technology in use; these include selective laser melting (SLM), electron beam melting (EBM), and direct metal laser sintering (DMLS).

## 3.3. Electronics 3D Printing

3D electronics printing utilizes various methods to print electronic and optoelectronic devices on various substrates like glass, plastic, and even paper. Here, the basic printing mechanics of layer-by-layer assembly still hold, but the additive materials include electronic inks. These inks contain minute particles of silver, copper, or other conductive materials, allowing the creation of circuits and other electronic components.

A crucial aspect of this process is the design file, which must account for circuit layout and component locations. Additionally, it must cater to the unique nature of 3D printing by ensuring traces are laid out in three dimensions, often around or even through the 3D printed object. Post-processing in electronic 3D printing is particularly important, as it usually involves 'curing' the electronic ink to guarantee conductivity.

## 3.4. Materials Used in 3D Printing

A critical aspect of any 3D printing endeavor is the choice of materials. In metal 3D printing, commonly used metals include steel, titanium, aluminium, and precious metals like gold and silver. These materials are usually provided as powders, with their fineness, grain shape, and quality significantly impacting the final product's properties.

In electronics 3D printing, the material palette includes conductive inks and dielectric inks. Conductive inks carry current and create the traces for the electronic circuits, while dielectric inks provide insulation between circuit layers. These inks' formulations are crucial; the particle size and type in the ink impact conductivity, flexibility, and durability of the printed circuits.

## 3.5. Advancements and Future Directions

3D printing technology is evolving at a breathtaking pace, with innovations arising in print speed, resolution, and novel materials. Simultaneously, multi-material 3D printers that can print both metal and electronic components within a single object are emerging, promising new opportunities for device integration and miniaturization.

The future may also see advancements in 4D printing or programmable materials. Here, the printed object can change its form, properties, or functionality after manufacturing, in response to external stimuli like temperature, light, or pressure.

In conclusion, the science behind 3D printing, specifically in metals and electronics, involves complex physical and chemical interactions, coupled with intricate design and process controls. However, the value proposition remains simple: unprecedented design freedom, customization, and integration. Indeed, this modern alchemy holds the potential to revolutionize how we create, making the impossible, possible.

# Chapter 4. Navigating the Material Universe: Metals in 3D Printing

In the land of 3D printing, the realm of metals represents a fascinating territory. Through this chapter, we will traverse this unique domain, unraveling the intricate intricacies of using metals in 3D printing. We start our exploration by comprehending the importance of metals, cutting across this vast material universe.

## 4.1. The Importance of Metals in 3D Printing

The importance of metals in 3D printing cannot be overstated. As we advance technologically, the demand for more robust, reliable, and complex components increases. Metals, with their high strength, toughness, and thermal resistance, naturally fit into these requirements. They engender parts and products of unmatched quality and longevity, essential in sectors where high performance is non-negotiable, like automotive, aerospace, and medical.

Metal 3D printing, otherwise known as Direct Metal Laser Sintering (DMLS) or Selective Laser Melting (SLM), uses a laser to fuse layers of metal powder together. This enables the creation of highly complex geometries that would be impossible or prohibitively costly with traditional methods, sounding the drums for an industrial revolution.

## 4.2. Understanding Types of Metals for 3D Printing

The next crucial phase in our journey is getting acquainted with the various types of metals used in 3D printing. Let's delve into some of the commonly used ones:

- **Aluminium**: Loved for its light weight and high strength, aluminium is used to make parts for aerospace and automotive applications, opening doors to greater fuel efficiency. Its high thermal conductivity and corrosion resistance further embellish its repertoire of utilities.

- **Titanium**: Celebrated for its superior strength-to-weight ratio, titanium finds ample use in fabricating parts for aeronautics and medical applications. Its innate biocompatibility allows it to be used for medical implants without causing adverse reactions.

- **Stainless Steel**: Known for its tenacity and resistance to corrosion, stainless steel is widely used to fabricate prototypes, tools, end-use parts, and jewelry.

- **Copper**: Highly conductive and corrosion-resistant, copper is primarily used for manufacturing heat exchangers, induction coils, and electrical components.

# 4.3. Selecting the Right Metal

Model selection in 3D metal printing is an art and a science. It's an art because, like a maestro selecting the right instrument for the right composition, you need to select the correct metal for your application. It's a science because the decision hinges on understanding the properties of the metal, such as strength, durability, heat conduction, electrical resistance, and cost-effectiveness. To simplify the procedure, create a checklist of key parameters relevant to your application, and pick the metal that most closely fulfills these needs.

# 4.4. Post-Processing and Finishing

3D printed metal parts often require additional processing post-printing. Post-processing encompasses support removal, strengthening treatments like heat treatment or HIP, and surface finishing methods such as bead blasting or polishing. Understand the post-processing requirements of your metal of choice, as this can significantly impact the total time, cost, and final properties of your 3D printed part.

# 4.5. Challenges and Opportunities

Like all technologies, 3D metal printing comes with its share of challenges - from the high cost of metal powders to the need for optimized printing parameters and advanced software solution necessitated by the complex geometries often associated with 3D printing.

However, the challenges are balanced by the enormous opportunities offered. From enabling bespoke manufacturing to weaving unprecedented intricacy into designs, from shortening product development cycles to reducing waste, 3D metal printing harbors potentials only limited by our imagination. As we deepen our understanding of this powerful technique and refine our capabilities, we can expect to unlock yet more of these potentials.

In conclusion, 3D metal printing is the new frontier of manufacturing. It offers a transformative way of designing and creating objects, conjuring complex and robust products in a way that aligns perfectly with the needs of our time. Our journey across the Material Universe remains incomplete without understanding and appreciating the powers and challenges of metals in 3D printing. As we continue exploring this innovative landscape, we become part of the alchemy that is redefining the world around us.

# Chapter 5. Electronics 3D Printing: Beyond the Silicon Age

The dawn of the third industrial revolution is hallmarked by the emergence of 3D printing technology - an innovation that has rewritten the rules of production and manufacturing across various sectors. One such domain where 3D printing has unleashed a new era is electronics. Referred to as "Electronics 3D Printing," the technology provides an imprint of the future, diversifying our imagination and practicability beyond the Silicon Age.

## 5.1. The Science: Understanding the Mechanism

Before embarking on an exhaustive exploration of electronics 3D printing, it is crucial to understand the underlying mechanism that helps turn designs into tangible outcomes. The method is fundamentally a fusion of additive manufacturing technology with micro-dispensing and micro-assembly techniques.

The process begins with the creation of a digital model that uses Computer-Aided Design (CAD) software. Specific components can either be designed from scratch or existing designs can be customized per requirements. The data from the CAD is then transformed into a format that the 3D printer can interpret and utilize.

During the printing process, the printer follows the CAD's instructions, releasing precise amounts of conductive inks through micro-nozzles across the X, Y, and Z axes onto a substrate. Layer upon layer, the printer builds the desired object, drawing from a variety of materials, from metals such as gold, silver, and copper to organic polymers and ceramics.

The high-level precision and control allow for the printing of intricate electronic components like layers of conductive traces and insulators. These can be integrated seamlessly into chips, circuits, and sensors, fundamentally transforming the way we fabricate and utilize electronic devices.

# 5.2. Evolution: Stepping into a New Era

Paved with the footsteps of innovation and driven by the ambition of limitless imagination, the journey of electronics 3D printing marks a significant turn from conventional and cumbersome manufacturing processes. As opposed to historical silicon-based electronics, this nascent technology allows for cost-effective, efficient, and rapid product development.

Traditional electronics manufacturing relies heavily on silicon due to its semiconducting properties. However, the process of turning silicon into wafers is arduous, costly, and fraught with environmental issues. Moreover, the physical properties of silicon limit the flexibility and miniaturization of electronics.

Electronics 3D printing, however, bypasses these constraints. It empowers designers to swiftly transition from prototyping to batch production, reducing the time-to-market significantly. It also enables complex geometries, facilitating the development of thin and flexible electronics that would be difficult to craft using traditional methods.

The shift towards this technology is not only industry-defining but also environmentally friendlier. The additive nature of 3D printing dramatically reduces waste generation, as only the required amount of material is used. The flexibility to use different materials also chips away at our reliance on silicon, which could deescalate the clash between technological advancement and environmental sustainability.

# 5.3. Promises and Possibilities: Envisioning the Future

As the global community continues to grapple with pressing issues like climate change, exponential population growth, and increasing demand for sophisticated electronic devices, the promise of 3D printed electronics seizes the spotlight.

To begin with, flexibility and customization are perhaps the most compelling attractions of this technology. From wearable tech to skin-like sensors, and from flexible displays to compact antennas, 3D printing satisfies the appetite for personalizing and miniaturizing electronic devices.

This advent also holds transformative implications for sectors like healthcare, where patient-specific electronic devices such as wearable sensors or bio-implantable devices can be developed. Likewise, in defence and aerospace sectors, lightweight and high-performance electronic systems, including UAVs and satellites, can be efficiently produced.

The democratization of production is another irresistible possibility. Given the right equipment and design files, anyone could print their electronic devices from the comfort of their own home. This transition could stir a paradigm shift, decentralizing manufacturing and enabling broad-based innovation in electronics.

In tandem with the Internet of Things (IoT), electronics 3D printing could play a pivotal role in realizing smart and connected cities. The potential for cheap, quickly produced and easily customized sensor networks and complex circuits can fast-track the implementation of

IoT solutions on a large scale.

# 5.4. Challenges and Roadblocks: Approaching with Caution

While the potential of electronics 3D printing is immense, it is crucial to note that the technology is still relatively young and presents its share of challenges.

Cost is a significant barrier, even though the technology promises to be more economical than traditional manufacturing in the long run. The initial investment in 3D printers capable of producing electronic components is currently relatively high, limiting its accessibility.

Substrate limitations and material considerations are other hurdles. The current 3D printers consume a significant amount of time to print multi-layer components as each layer needs to dry completely before the subsequent layer can be added.

Integration with traditional manufacturing processes may also pose significant hurdles, as companies will have to overhaul established systems and retrain employees to adapt to new processes.

Intellectual property issues may arise as 3D printing technology makes replicating designs much easier, potentially leading to patent infringements and other legal complications.

# 5.5. Conclusion: The Road Ahead

The journey of electronics 3D printing is just beginning, and it's a road filled with promises and potential. Given the pace of technological evolution, we may very well see this disruptive technology overcome its challenges and touch lives in ways previously unimaginable. The exciting possibility of a world where we can print circuitry as easily as we print paper suggests the dawn of a new era in electronics manufacturing. It makes us eager witnesses to the evolution of what could be the next Silicon Age – a revolution orchestrated by the magic of modern alchemy that is 3D printing technology.

Yes, there are hurdles to cross, but faced with the potential rewards, the case for electronics 3D printing remains robust. As we stand on the threshold of this new dawn, one thing is sure: the journey has just begun, and it promises to be an exciting one.

# Chapter 6. Unraveling the Process: How 3D Metal Printing Works

Subsequent to the invention of 3D printing in the early 1980s, the technology has made breathtaking strides, notably in the field of metal fabrication. The process involving additive manufacturing and metallic materials is often known as 3D Metal Printing.

## 6.1. Basics of 3D Metal Printing

Being a form of additive manufacturing, 3D metal printing primarily relies on the principle of layering. It essentially involves depositing thin layers of metal powder and selectively fusing them together, layer by layer. This method is possible for a wide array of metals and metal alloys. However, the strength, durability, and characteristics of the final product can vary significantly by the metal used and the layering technique employed.

The road to a 3D metal print commences with a virtual blueprint. This blueprint, or 3D model, is technically developed using CAD (Computer-Aided Design) software. The CAD data is then processed by slicing software that splits the model into several thin layers. The slicing software generates specific instructions to the 3D printing machine about how each layer needs to be printed.

Another significant term in the realm of 3D metal printing is "supports." Throughout the printing process, there might be sections of the model that are overhanging or structurally complex. These sections require additional support during printing, which are correspondingly added to the CAD model and later removed after printing.

# 6.2. Technologies in Metal Printing: Powder Bed Fusion and Directed Energy Deposition

The cosmos of 3D metal printing is wide and comes equipped with several techniques. Two major methods have garnered significant attention and widespread usage: Powder Bed Fusion and Directed Energy Deposition.

Powder Bed Fusion (PBF) includes methods like Selective Laser Melting (SLM), Direct Metal Laser Sintering (DMLS), and Electron Beam Melting (EBM). The PBF process begins with spreading a thin layer of metal powder (20-50 micron range) onto a build platform. A high power laser or an electron beam then selectively fuses the regions of layer based on the CAD data. The build platform is then lowered, another layer of powder is applied, and the process continues until the part is fully formed.

In contrast, Directed Energy Deposition (DED) is a process more akin to welding. This technique uses a nozzle to deposit metal powder or wire into a melt pool created by a laser or an electron beam. The melted material solidifies to form a layer, and subsequently additional layers are deposited on top. Comparatively, DED enables large-scale applications and repairs of existing parts more conveniently than PBF.

# 6.3. Examining Key Parameters in 3D Metal Printing

Several parameters are influential in the final outcome of 3D metal printed parts. This includes power and speed of the applied energy source, layer thickness of the powder, and the thermal properties of the metal powder. The interplay of these factors critically determines the mechanical properties, strength, resolution, and even potential defects of the printed object.

The power and speed of the laser or electron beam influence how well the metal particles fuse together and the cooling rate, which has implications on the microstructure and resultant properties of the metal object. Thin layers might be more time-consuming to process compared to thicker layers, but they often offer better resolution and surface finish.

# 6.4. Future Perspectives: Tailoring Properties and Multi-Metal Printing

With 3D metal printing scheduled for the next phase of progression, one fascinating avenue is tailoring the metal's properties. This reflects the capability to control the microstructure and hence, the mechanical properties of the printed metal part at a granular level. This would involve computationally predicting the system's thermodynamics and kinetics and optimizing the operating parameters accordingly.

Another potent prospect is multi-metal printing. Imagine being able to create objects with multiple metals, each with its unique properties and functionally graded. Indeed, this emerges as a significant prospect in the coming age, making 3D metal printing an alchemical process of our times.

In conclusion, 3D metal printing represents an extraordinary tool in the hands of engineers and designers today. Keeping abreast with the nuances of its functioning and unleashing its power has incredible potential. By navigating the complexities, we are tapping into a realm that can skillfully orchestrate form, function, and material properties in unison, thus opening remarkable avenues for innovation and growth.

In the realm of modern alchemy, understanding 3D metal printing serves as a key entry point. Hence, it is crucial to keep exploring, learning, and unlocking the possible marvels that the technology promises. The opportunities are infinite; all it requires is imaginative zeal and a pursuit for understanding the craft that could significantly redefine our future.

# Chapter 7. Printing Circuitry: The Future of Electronics Manufacturing

The dramatic transformation of electronics manufacturing is becoming more tangible as we steer towards an era where circuitry can be printed through the marvel of 3D printing technologies - a significant leap that promises to revolutionize the industry. This promising future encompasses numerous advantages, including reduced manufacturing costs, faster prototyping, and even the ability to create more compact and complex devices.

## 7.1. Rapid Prototyping and Customization

3D printing allows the creation of circuitry in just a few hours compared to the traditional circuit manufacturing methods that could take weeks. This vast reduction in lead times enables quicker iterations, facilitating the testing of different designs, and significantly speeding up the process to arrive at an optimal solution. Such rapid prototyping also expands the potential for customization. Today, we dream of gadgets persistently powered by solar panels printed directly onto their surfaces, headphones with integrated circuitry, and eyewear lenses doubling as screens. These possibilities might seem far-fetched, but such customization is within our grasp thanks to the advances in 3D printed circuitry.

## 7.2. Cost Efficiency and Scalability

The financial implications of 3D printed electronics are equally compelling. Traditional circuit board production is laden with substantial fixed and labor costs, particularly for small production runs. In contrast, 3D printing brings forth a new, cost-effective approach. The availability of affordable 3D printers and the absence of costly molds mean shorter production runs can be carried out inexpensively. At the same time, the scalability potential of 3D printing plays a crucial role. It is a technology that can flexibly cater to both low-volume, high-customization orders and high-volume, low-customization batches with equal ease.

## 7.3. Miniaturization and Complexity

3D printed circuitry paves the route to miniaturization and enhanced complexity. This technology enables electronics with incredibly small footprints, including nanoscale circuitry. As we strive to make devices more compact and functional, such miniaturization takes paramount significance. Moreover, 3D printing allows us to break away from the planar confinement of traditional circuit boards. We can create multi-layered circuitry, overcoming the traditional two-dimensional limitations, opening doors to new possibilities in electronic design.

## 7.4. Material Science in Printed Electronics

Material science plays a significant role in printed electronics. Currently, silver nanoparticle ink is the most widely used due to its excellent electrical conductivity and compatibility with inkjet printing. Advanced research is now being conducted with graphene and other conductive polymers to drive future advancements in 3D printed electronics.

# 7.5. Environmental Benefits of 3D Printed Electronics

With its additive manufacturing process, 3D printing significantly reduces material wastage compared to subtractive traditional manufacturing practices. Besides, recycling could be more achievable with certain types of thermoplastic materials, contributing positively to sustainability goals. Nevertheless, it's important to note that despite these potential benefits, there are environmental implications associated with the energy consumption of 3D printers and the lifecycle of the materials used.

# 7.6. Bridging the Gap: Challenges and Solutions

While 3D printed electronics hold remarkable promise, we face associated challenges. For instance, the conductivity of printed materials often lags behind the performance of traditionally manufactured counterparts. However, advanced material research and improved printing resolution are bringing solutions closer.

Apart from these technological hurdles, there are regulatory challenges, like safety standards and intellectual property protection, which require careful addressal. Yet with evolving norms and international cooperation, these challenges can be appropriately managed, paving a path for the bright future of 3D printed electronics.

So, let's embark on the remarkable journey of 3D printed electronics together, embracing and overcoming challenges, as we unlock the keys to a revolutionary future in electronic manufacturing filled with unimaginable opportunities.

# Chapter 8. Notable Case Studies: Success Stories in 3D Metal and Electronics Printing

Though relatively nascent, the field of 3D metal and electronics printing has witnessed a surge in a plethora of remarkable innovations, showcasing not just its potential but its maturity as a viable manufacturing tool. Let's delve into an analysis of a few pioneering revolutions, each demonstrating a different aspect of the far-reaching applicability of 3D printing in both metals and electronics.

## 8.1. GE Aviation: The LEAP Engine

In the realm of aero-engine manufacturing, GE Aviation has left an indelible mark in recent history, courtesy of its LEAP (Leading Edge Aviation Propulsion) engine. This high-pressure turbine part was crafted using a direct metal laser melting (DMLM) technique, which vastly differs from traditional casting methods. By integrating 3D printing, GE was able to reduce the part's weight by as much as 25%, increase fuel efficiency by 15%, and significantly trim manufacturing time. This unprecedented feat pushed the boundaries of industry standards, and today, thousands of LEAP engines are in service worldwide.

## 8.2. SpaceX: SuperDraco Engine Chamber

Space exploration suggests the ultimate combination of precision, reliability, and advanced technology application. Space transport services company SpaceX received widespread recognition, especially for its utilization of 3D printing. Among their myriad accomplishments, the SuperDraco is a remarkable one. This reusable thruster, extensively made through direct metal laser sintering (DMLS), propels the Dragon vehicle's launch escape system, highlighting its paramount importance for the safe abort function during a manned mission. SpaceX's success serves as a testament to 3D printing's potential in delivering complex structurally sound components for space applications.

## 8.3. Siemens: 3D-printed Gas Burners for Turbines

Harnessing 3D printing's ability to manufacture custom parts swiftly, Siemens 3D-printed its first gas turbine blades. Not only did the process cut down on the lengthy production cycle — from two years to just eighteen months — it also resulted in fuel-efficient, robust parts able to withstand high pressure and temperature extremes. Siemens' gas turbine case establishes that component-level innovation empowered by 3D printing can significantly boost operation efficiency in energy sectors.

## 8.4. Ottobock: Myoelectric Prosthetics

Another area where 3D printing has made a significant stride is in the healthcare sector. Ottobock, a German prosthetics company, has been utilizing 3D printing to create accurate, highly functional myoelectric prosthetic hands — a feat that would be daunting with traditional manufacturing methods. This trailblazing innovation has resulted in lightweight, cost-effective, and user-responsive alternatives, a lifeline that offers newfound freedom and dexterity to amputees.

## 8.5. Nano Dimension: 3D-printed Electronics

Success stories of 3D printing in electronics may not be as copious as in metal, but Nano Dimension uniquely stands out. The Israeli company revolutionized rapid prototyping with its DragonFly 2020 Pro 3D printer, capable of printing multi-layer circuit boards. This capability not only accelerates the development cycle but also drastically reduces the costs associated with creating prototypes for complex boards. Nano Dimension's innovation underlines 3D printing's potential as a game-changer in electronics prototyping and production.

## 8.6. Lockheed Martin: Spacecraft Fuel Tanks

In the domain of spacecraft manufacturing, Lockheed Martin has successfully 3D printed two large titanium fuel tank prototypes for satellites. The use of 3D printing enabled the manufacture of the tanks as a single piece, bypassing the several hundred welds typically required, reducing cost, and enhancing reliability. This accomplishment further establishes the efficacy of 3D metal printing in producing large, high-quality components in demanding sectors such as aerospace.

Each of these success stories underscores how 3D metal and electronics printing has crossed the chasm of uncertainty to render itself an invaluable tool for various niches, industries, and applications. These entities are not solely heralding entirely new possibilities in design and manufacturing but also redefining conventional wisdom that has gripped respective sectors. As more areas begin to discover and unlock 3D printing's potential, we can expect even more inspiring innovations in the future.

# Chapter 9. The Challenges and Solutions: Overcoming Barriers in 3D Printing

Despite the significant advancements in 3D printing, particularly in metal and electronics, the technology faces several challenges that currently limit its potential. This segment aims to explore these challenges as well as suggest suitable solutions that could enhance the usability, efficiency, and accessibility of 3D printing technologies.

## 9.1. The Challenge of Post-Production Processes

One of the primary challenges of 3D printing in metals and electronics is the extensive post-production process often associated, which can include heat treatment, surface refinement, and quality assurance. Post-production processes may cause delays, increase costs, and exert a considerable strain on resource allocation.

The reliance on these ancillary procedures prevents 3D printing from fulfilling its potential as a standalone fabrication method. Complicated geometries and internal structures that are easily achievable in the 3D modeling phase can become bottlenecks in post-production, adding to the duration and complexity of production.

To counter this issue, we must develop and incorporate more advanced support structure design tools and material-specific printing parameters. By improving the precision of support structures and the conditioning of materials during the printing process, it should be possible to minimize the amount of post-production work required. Moreover, implementing in-line quality assurance tools could potentially aid in reducing the post-production process.

## 9.2. High Material Costs

Material cost is another fundamental obstacle faced in 3D printing, especially when it comes to metals. The material costs can contribute significantly to the overall cost of production, thus limiting the adoption of this technology in industries where profit margins are thin.

To overcome this hurdle, the concept of material recycling should be introduced in the workflow, wherever applicable. Using sophisticated technologies like selective laser sintering, unused materials can be reused in subsequent printing processes, reducing waste and hence, the overall cost. Additionally, partnerships with metal suppliers to reduce bulk costs could be another effective measure.

## 9.3. Incomplete Fusion and Inconsistent Properties

Incomplete fusion or other defects during the 3D printing process can result in products with inconsistent properties, compromising the product quality. These inconsistencies could be a result of variation in process parameters or raw material characteristics.

Addressing this issue requires an approach that combines improved material handling, precision engineering, and quality assurance protocols. Real-time monitoring and control of the 3D printing process could identify and rectify inconsistencies as they occur, significantly reducing product failure rates.

## 9.4. Limited Material Choices

The use of 3D printing in metals and electronics usually requires specialized materials that respond well to the production process. This requirement limits the choice of materials that can be used, often excluding traditional materials used in standard fabrication processes.

The R&D approaches taken in the 3D printing industry are already trying to expand upon the available material choices. Scientist and engineers are developing new metal alloys specifically formulated for 3D printing processes, which could dramatically enhance the range of material choices and the versatility of applications.

# 9.5. Lack of Standardization

The absence of standardization for 3D printing processes, materials, and end-product testing causes significant hurdles, particularly for industries regulated by quality control standards and guidelines.

Creating universal benchmarks for 3D printing is crucial. Collaborative efforts between industry leaders, researchers, and standards organizations are necessary to form comprehensive, globally accepted standards for 3D printing.

# 9.6. Intellectual Property Concerns

3D printing allows nearly anyone with adequate equipment to produce virtually anything, leading to concerns about intellectual property rights infringements.

Regulations need to be implemented or adapted to manage intellectual property rights within the context of 3D printing. The laws will need to adapt to the changing nature of production and distribution resulting from this technology.

The challenges highlighted above represent some of the broader issues affecting the adoption of 3D printing in metals and electronics.

However, these challenges are not insurmountable. Through continued research, technical advancements, and policy-making efforts it is feasible to envisage a future where these barriers are suitably negotiated. The potential held by 3D printing is enormous; it is only a question of how quickly we can adapt and grow to reap its benefits.

# Chapter 10. Environmental, Health, and Safety Considerations in 3D Metal and Electronics Printing

While 3D printing holds vast potential for revolutionizing metal and electronic fabrications, it is also important to spotlight the environmental, health, and safety (EHS) considerations that come along with progress. Striving for clean, safe, and sustainable advancements, the following section provides comprehensive guidance on EHS aspects associated with 3D printing in metals and electronics.

## 10.1. Occupational Hazards in 3D Metal and Electronic Printing

Certain occupational hazards are inherent to 3D printing that requires attention. Metal and plastic dust particles, known as particulate matter, can pose severe health risks when inhaled or come into contact with skin. Workers exposed to printing materials such as metals, resins, or polymers could develop allergies, skin irritation, or respiratory diseases. To mitigate these risks, it is essential to maintain good ventilation, utilize personal protective equipment (PPE), and establish regular health surveillance for exposed workers.

Additionally, the high-temperature processes involved in metal printing can lead to burns and injuries. Measures should be in place to prevent direct contact with hot surfaces. Providing training, using appropriate safety gloves, and adhering to lockout/tagout procedures are some methods to counter these risks.

## 10.2. The Impact on the Environment

3D printing, like any manufacturing process, has an environmental footprint. It can result in the emission of potentially harmful gases and particulates, contributing to air pollution. Furthermore, the use of energy-intensive processes and non-recyclable materials can significantly impact climate change and ecological health.

Companies should work towards minimizing these impacts in manufacturing processes by adopting renewable energy sources, improving printer efficiency, and promoting the recycling of printing materials.

## 10.3. Material Safety and Handling

Material safety and handling is another significant EHS consideration for 3D printing. Both raw materials and printed parts can present hazards, such as reactive chemicals or heavy metals. Employees must undergo thorough training on safe handling procedures, proper storage, and disposal techniques. The use of material safety data sheets (MSDS) and adherence to regulations regarding hazardous substances can aid in minimizing associated risks.

## 10.4. Facilities and Equipment Safety

Maintaining the safety of facilities and equipment is crucial in a 3D printing environment. Adequate preventive maintenance of machines can prevent malfunctions and accidents. Similarly, facilities must be designed keeping in mind fire safety, emergency response, good ventilation, and ergonomic considerations.

## 10.5. Regulatory Compliance

Compliance with local, national, and international regulations is a requisite aspect of EHS management in 3D printing. Compliance activities can range from maintaining air emission permits, hazardous waste disposal certifications, to workplace safety oversight. Clear understanding and adherence to these regulations can ensure that the implementation of 3D printing technologies does not compromise environmental sustainability, worker safety, or community health.

## 10.6. The Role of EHS Management System

An effective EHS management system is instrumental in recognizing, controlling, and mitigating environmental, health, and safety risks in 3D printing. The EHS management system should encompass comprehensive risk assessment, safety policies and procedures, regular EHS training and awareness programs, audits, and continual improvement initiatives.

## 10.7. Looking Forward: Opportunities for Sustainable 3D Printing

The 3D printing industry also presents opportunities for environmental sustainability. We envision a future where 'green' printing materials become the norm, and recycling and upcycling of waste from 3D printing are standard practices. Energy-efficient printers and the adoption of renewable energy sources can also help curb carbon emissions.

By balancing innovation with EHS responsibility, we can harness the power of 3D printing to transform our society, economy, and environment sustainably. This synergy of progress and responsibility is the essence of modern alchemy.

This comprehensive aspect of EHS considerations in the world of 3D printing serves as a direction and guide to anyone involved in this revolutionary technology, setting the pathway to a responsible and sustainable future. As we continue our exploration into the fascinating realm of modern alchemy via 3D printing, let's ensure to uphold our commitment towards environmental sustainability, personal health, and safety.

# Chapter 11. Looking Ahead: The Future Landscape of 3D Printing in Metals and Electronics

As we stand on the precipice of technological developments, the expanding frontiers of 3D printing have been mapping changes of unprecedented magnitudes. The domains of metals and electronics have been the focus of these remarkable shifts, bringing ideas forth from the realm of imagination and materializing them in the physical world.

## 11.1. Reshaping the Manufacturing Landscape

The power to manufacture complex geometries where traditional machining fails, coupled with the capability for high repeatability, gives 3D printing an edge in the world of manufacturing. This paradigm shift points towards an ever-increasing adoption of 3D printing in metals and electronics. It has already distorted traditional production chains, transitioning from a prototyping tool to a fully integrated manufacturing solution.

Benefits like waste reduction and quick turnaround times have been pivotal in empowering industries to optimize their manufacturing chains. The aerospace and automotive sectors have especially thrived, allowing them to produce lightweight components and complex engine parts. Similarly, the medical industry leverages this unique fabrication process to create personalized implants and surgical instruments, providing greater patient care.

As technology advances, the development of multi-material 3D printing, offering the flexibility of combining varied materials in a single printing process, could redefine versatility in manufacturing. Customized composite materials, as a result, could elevate the strength, flexibility, and utility of 3D printed products significantly.

## 11.2. Electrifying Progress in Electronics

The field of 3D printed electronics has been witnessing a surge of novel methodologies that might revolutionize traditional circuit board production. The advent of conductive inks and filament has enabled the fabrication of intricate circuits directly into a variety of substrates. It paves the way for the development of truly integrated electronics – tighter, lighter, and more flexible than ever before.

This could segue into a new dimension of product design and functionality, prompting devices to become increasingly compact and integrated. Advances in nanotechnology supplement this trend towards miniaturization in electronics, pushing the boundaries of what's possible within circuitry. Wearable technology, such as smart textiles and medical implants, stands to benefit considerably from this convergence of sophistication, integration, and size reduction.

Moreover, the debut of 4D printing, where the printed object can change form or function over time in response to environmental factors, holds significant implications for electronics. This development could fuel the creation of intelligent, adaptive devices with self-healing capabilities, dramatically improving product lifespan and reliability.

## 11.3. A Sustainable Future

3D printing processes in metals and electronics have the potential to be incredibly sustainable. The transition from subtractive manufacturing to additive reduces the volume of raw material waste significantly, and novel methods like direct energy deposition promise further sustainability with near-net-shape output.

Moreover, the use of recycled materials in 3D printing has gained traction. The circular economy, an economic system aimed at eliminating waste and creating a continuous loop of resource reuse, aligns naturally with 3D printing. By converting scrap metal into valuable resources, industries can minimize waste generation, lower raw material costs, and reduce their carbon footprint.

In a step towards energy efficiency, researchers are exploring new means of power optimization during the 3D printing process. Controlled pulsing of print lasers, for instance, can noticeably reduce energy consumption while maintaining product quality.

# 11.4. Challenges Ahead

Despite the many exciting prospects, the path to large-scale adoption is not without hurdles. The cost of metallic 3D print materials remains prohibitive for many potential applications, and the higher cost of energy and slower production rates, compared to traditional manufacturing methods, limit its appeal to certain industry sectors.

Quality control is another concern. Despite the geometric freedom and customization capabilities, variation in product quality often arises, leading to implementation uncertainties. Rigorous testing and inspection procedures, as well as advances in Artificial Intelligence and machine learning in process monitoring, could mitigate these concerns.

Further, the design paradigms for conventional manufacturing do not always translate well into 3D printing. Up-skilling the existing workforce to harness the full potential of this technology and encouraging a shift in design thinking are vital.

# 11.5. Final Thoughts

The possibilities offered by 3D printing in metals and electronics are expansive, from revolutionizing manufacturing chains to fostering ever-greater intricacy in product design. It has the potential to redefine how we perceive material utility, disrupt traditional production paradigms, and carve a sustainable path ahead.

As we advance deeper into the territory of modern alchemy, the journey will undoubtedly be challenging. Nonetheless, it is a challenge imbued with immense potential for our world's transformation. Weaved into this tapestry of changes, the tales of 3D printing continue to unfold, promising to create a future akin to what once belonged in the realms of science fiction.

www.ingramcontent.com/pod-product-compliance
Lightning Source LLC
Chambersburg PA
CBHW071049260726
48661CB00007B/3225